Nigeria's Adventures in Technology

Dike N Kalu

Outskirts Press, Inc.
http://www.outskirtspress.com

ISBN: 978-1-4787-6140-2

The Nigerian edition was published in 2014 by SICA GRAPHICS AND PUBLISHING Onitsha, Nigeria.

PRINTED IN THE UNITED STATES OF AMERICA

Dedication

To my adopted student body of Virtual University without Walls and the Nigerian public, especially the youths.

And to the memory of my good brother and friend Otisi J. N. Kalu (alias Abbott), Achi Na Agbara Oha of Abiriba - an illustrous son of NdukwoNta.

Contents

Acknowledgments

I thank Chief Sam Amuka and the Nigerian Vanguard Newspapers for kindly publishing the essays in this book at the time they were written. Although they've now been thoroughly revised, the fact that the substance of each essay was not altered is evidence that the issues raised when the essays were first written are still very much alive today. I also thank my friend, Ole Uduma Ikpa for reading the final draft of this book and making many helpful suggestions.

I thank Professor Anya for writing the Preface for this book at a very short notice, and the Nigerian edition's publisher Chidi Ikenna Amobi and his team who worked tirelessly to meet publication deadlines.

Finally, I thank all the faceless and dedicated people that have made the Internet such an invaluable source of information for the discerning public.

http://www.profdikebooks.com

Also by Dike N Kalu

Agony of a Nigerian Scientist

Science and Technology in Nigeria

Preface

This slim volume, Nigeria's Adventures in Technology by one of Nigeria's most accomplished and dedicated scientists is an important and timely piece of advocacy for the centrality of science and technology in the affairs of a nation. It achieves the important objective of popularizing science and technology by capturing the human imagination. He achieves this by illustrating the achievements of Nigeria and Nigerians through our recent dalliance with the high technologies of space science by launching space satellites and pursuing the development of critical capabilities in nuclear science. In doing this, he unintentionally emphasizes a point made many years ago by the great Pakistani Physicist and Nobel laureate Abdus Salam that the fastest way for the developing nations to leapfrog into modern science and technology is by developing capacity in the higher technologies typified by space science and nuclear technology.

You may ask, what is the point in all these? It is to illustrate the simple fact that in the modern world no nation that did not embrace or commit to the centrality of science and technology in its development has been able to tackle the challenge of development. Examples abound: United States, South Korea, and now China, Brazil and India. It is the important point that I attempted to demonstrate in my book, Science and the Crises in African Development (1993). The failure of African nations to tackle the challenge of underdevelopment through the instrumentality of science and technology has been a sour point particularly in the realm of public policy. Professor Dike N Kalu subtly makes this point by not only advocating the need for a revolution in science and technology in Nigeria but particularly by illustrating the relevance of science and technology to national development and on a nation's quality of life through their impact on life expectancy. Noteworthy is his brilliant strategy of adopting the conversational format of dialogue with (his) students to make his points. It is simple, imaginative and effective.

There is an important and forceful point implicit in Professor Kalu's advocacy which can easily be glossed over: that the impact of science and technology in the development of a nation cannot be fully harnessed unless the assumption and methodologies of S & T are embedded in the culture. Thus, the scientification of the culture of the Nigerian peoples is the most central challenge facing us in all our endeavors to emerge as a powerful economy in the 21st century. In this effort, engaging the youth, the politicians and the public policy mandarins has proved a major obstacle. Professor Kalu's simple but popular engagement with the Nigerian population is therefore an important starting point.

This slim volume is destined to make an impact beyond its size by the revolutionary approach of its engagement with the Nigerian peoples especially the youth. Professor Kalu deserves our congratulations and appreciation for this potentially transformative effort for Nigerian development and for the Nigerian polity.

Professor Anya O. Anya, Ph.D. (Cambridge), D.Sc. (Hons), FAS, NNOM; Past President, The Nigeria Academy of Science and Past Chairman, Governing Board, Nigeria National Merit Award.

Introduction

First and foremost, it is important that the reader be aware of the reason for writing the essays in this book in order to fully appreciate them.

Nigeria got political independence in October 1, 1960 and the author, Dike N Kalu was among the students that went abroad the following year to obtain higher education and return to Nigeria to contribute in the building of their new nation.

Dike N Kalu got his first training in Medical Laboratory Technology in London (1961-1964) before going to Queen Elizabeth College, London University, U.K. where he obtained his Bachelor's degree in 1968. In 1971 he got his Ph.D. in Biochemistry and Zoology from the same university, and in the same year he was elected Fellow of the Institute of Medical Laboratory Technology (FIMLT) in the United Kingdom.

Dr. Kalu was ready to return home to Nigeria with his family but an in-law who was a journalist in Nigeria at the time advised him not to return yet because things were still unstable at home as the Nigeria-Biafra war had just ended.

As a result the author took his family to the United States of America where he seized the opportunity to extend his studies as a postdoctoral fellow in Endocrine Physiology at the Johns Hopkins University School of Medicine, Baltimore, Maryland.

In 1975 Dr. Kalu was offered a faculty position as Assistant Professor of Physiology at the University of Texas Health Science Center, San Antonio, Texas, where he taught endocrine physiology to graduate, medical, and sometimes dental students, and did basic and applied research. He climbed the academic lather and was made full professor in 1986.

In 1998 the University of London awarded Professor Kalu the D.Sc. degree for his contributions to research in Calcium and Bone Metabolism. When he retired in 2002 the University of Texas Health Science Center, San Antonio honored him with the award of Professor Emeritus.

Although Professor Dike, as he is fondly known, appeared to have all the outward trimmings of a successful academic career he always felt that something was missing in his life.

He was uncomfortable with the fact that there were no other Nigerians doing substantive work in his chosen field of Calcium and Bone Metabolism. He knew because he constantly read the relevant literature, wrote extensively on the subjects, and he often attended national and international meetings on Calcium and Bone Metabolism, and Nigeria was never represented.

By August 1974, Dr. Kalu had started applying to universities in Nigeria for a faculty position so that he could go home and join his peers to contribute in the training of Nigerian biomedical students and in advancing science in the country through basic and applied research.

In 1984 Dr. Kalu made some fateful decisions. He slowed down his research at his university in the United States, stopped writing grants for research support, suspended taking graduate and post doctorate students into his well-funded laboratories and refused appointments to departmental, university and national scientific society committees. He deemed all these as necessary in order to ensure that he was ready to leave his faculty position in the United States and return to Nigeria at a moment's notice.

In connection with his quest for a faculty position in Nigeria, Dr. Kalu visited the country many times; sometimes just to make doubly sure that his hand-carried applications did arrive at the universities since he received no acknowledgments. Occasionally he combined his self-initiated university interviews with a lecture/seminar to university students.

One of the most fulfilling moments of his academic life was when he gave a lecture/seminar on "Calcium Metabolism" to a sea of black students at the University of Port Harcourt, the first time in his life that his audience wasn't predominantly white!

During his visits to Nigerian Universities, he always received favorable remarks from the faculties. Some said his credentials were outstanding, and many admired his accomplishments and his publications of over 200 scientific articles about 100 of which are in peer-reviewed international journals. But none of the six Nigerian universities he had applied to recruited him after over ten years of trying.

Professor Dike published his unfortunate experiences with the Nigerian University System in a book he titled "The Agony of a Nigerian Scientist" (Heinemann Educational Books, Nigeria, Plc, 2003). He found writing this book cathartic and it was only after he wrote it that he could summon up courage to pick up the pieces and resume his academic life as a reluctant exile in the United States.

Having been rejected by his home universities, Professor Dike resolved that henceforth the Nigerian nation would become his Virtual University Without Walls and every Nigerian that so wished would be his student. Periodically, he would communicate with his adopted student body on important issues concerning Nigeria through the writing of essays on the subjects.

When Professor Dike was asked why he was assuming such a demanding task of making the Nigerian public his student body solely at his own expense, he replied:

"Whenever fate puts one in a position of service to society due to a unique talent or training, failure to serve and serve honorably is a violation of moral responsibility".

Professor Dike's arduous undertaking was eased by his good friend Chief Sam Amuka, the publisher of Vanguard Newspapers who in effect cheerfully undertook the burden of disseminating his writings to the Nigerian public in the Vanguard.

As Professor Dike's University activities in the US re-blossomed he once again became mentor to many American postgraduate students and to postdoctoral fellows from Sudan, India, China, Iran and Nigeria. The Nigerian was later appointed Director General of NAFDAC in January 2009.

So the essays that follow were written in accord with Professor Dike's wish to communicate periodically with his adopted students through publications on important issues that affect Nigeria. This little book of essays is dedicated to them and the Nigerian public especially the youths who are the inevitable leaders of the future.

The first essay in this publication, Revolution in Science and Technology in Nigeria, deals with why these enterprises have failed to thrive in our country. This is of overwhelming importance since most Nigerians had hoped that successive administrations after political independence in October 1960 would use modern technology to advance socioeconomic development in Nigeria. As a result of Nigeria's recent venture into satellites, Chapter 2 deals with Satellites and Shuttles, which the author believes are technological marvels. A discussion of Nuclear Energy in Nigeria follows in Chapter 3, and highlights the fact that while it is desirable for a country to strive for energy sufficiency one has to be mindful of the deleterious consequences of some technologies. The last chapter examines Life Expectancy in Nigerians which should be of interest to all our people in view of the abysmally short length of life in the country

Each essay in this book underlines in its own way the necessity for indigenous competence in science, technology and education in general in Nigeria.

CHAPTER **One**

Revolution in science and technology in Nigeria

THE ESSAYS IN this book continue my commitment to make my student body and the Nigerian public better informed about important issues that affect our lives. The challenge is to treat a subject matter in a way to accommodate our wide and diverse backgrounds.

Because of the acknowledged importance of science and technology in social and economic development, this first essay will examine some of the factors necessary for science and technology to thrive in Nigeria. We will begin with the basics of the subject matter.

Science is essentially knowledge of how the natural world works.

Consider for a moment the fact that many natural phenomena occur with such regularity that they are so often taken for granted. For instance, as you read this chapter in a relatively unexcited state, every minute you breathe about 15 times and your heart makes about 70 beats. Now that I have alerted you to these facts, you can probably feel your heart beating, and you can even consciously modulate the frequency of your breathing. It is not intuitively obvious why we breathe and why our heart beats all the time, and what control the regularity of these events are hardly the subject of normal conversation.

Similarly it is not usual for us to stop whatever we're doing to think about why we wake up with such regularity each morning as the sun rises in the East and we end each day as the sun disappears from the horizon in the West at a point diametrically opposite to where it rose from.

We are all familiar with the saying that what goes up must come down. Accordingly any time you throw up an object such as this book, it always falls back to the ground. The same thing will happen to every object thrown up by anybody in any corner of our planet. And yet it is not instinctively obvious why these objects fall back to the ground.

Finally, consider that while nobody can predict with any degree of certainty what lies ahead for any baby, we know for certain that barring an unfortunate accident or a fatal illness a baby will grow up into a young person, mature into an adult, grow old and eventually die. We do not normally stop in the middle of our daily tasks to ponder how and why human beings and for that matter all living things age. We take it for granted; it is natural.

However, there is a group of people whose vocation is to try to unravel the mysteries of nature. These individuals are characterized by insatiable curiosity and the urge to explore and discover the secrets of nature. They ask questions about why nature is the way it is. They never take things for granted, and they try to discover the reasons behind such natural phenomena as we alluded to above by an ordered methodological approach. These are the scientists, and their vocation is science.

Scientific knowledge is best acquired by experimentation and observation in the controlled environment of a laboratory. From experiments, scientists make observations and these observations form the basis upon which they construct theories about nature. The progression from the collection of observable information or data to the erection of a theory or law of nature based on these data is called "the principle of induction", and it is the basis of modern scientific research.

While a scientist is an individual engaged in the business of acquiring new knowledge, a technologist is involved in applying knowledge to provide practical tasks. Societal advances largely depend on man's ability to transform knowledge into technologies such as telephones, cell phones, different forms of energy, TVs, aircrafts, medical technologies such as X-rays and MRIs, electric lights, computers, the Internet, satellites, etc. There is virtually no aspect of human existence that is not impacted by the use of technology. In fact all underdeveloped countries are characterized by their failure or inability to use the powers of technology to solve their developmental needs.

A technologist can also be a scientist, and vice versa. Both are necessary in any society.

It is obvious that science and technology are two words joined together by the conjunction "and". They are often used together because as we have seen, they are closely related. In fact some treat science and technology as a singular word as I have done in this book. The popularity of the term science and technology derives from the fact that human civilization practically depends on it. Worldwide, the use of science and technology has proven to have a good track record in enhancing the development of countries.

I subscribe to the view that if one is convinced of the power of science and technology in promoting development as I am, one is then obligated to continue to implore the Nigerian government and the public to do what advanced nations do regularly. Advanced nations have, for centuries, continued to translate science into productive technologies and use the latter to fuel the development of their countries.

Nehru, the late prime minister of India could not have put it more succinctly when he said:

"I do not see any way out of our vicious circle of poverty except by utilizing the new sources of power which science has put at our disposal".

At the start of the scientific era, nature was studied by trial and error using natural ingenuity and very simple instruments and methods.

Even though unsophisticated techniques were used in the early experiments done centuries ago, the conclusions that were drawn from them are still valid today. This is because science utilizes very stringent methods and demands that scientists' observations must be repeatable by anyone in any corner of the earth.

In other words, experimental observations made by scientists are universal phenomena and can be treated as basic facts.

Types of Scientific Research: Scientific research can be classified under three headings, (a) basic or fundamental research, (b) applied research, and (c) developmental research.

Basic research is the search for truth about nature for its own sake. Its intent is not to create or invent a product. The sole aim of basic research is to contribute to the pool of new knowledge and thereby provide a better understanding of the subject being studied.

One example of basic research is the research that led Watson and Crick to describe accurately in 1953 the helical structure of DNA, an intracellular macromolecule that carries genetic information with precise instructions for the development and physiology of living things including humans. Basic research is very expensive and is usually funded by the government.

In contrast to basic research, applied research is usually carried out to address a specific problem and it leads to the invention of products or services or to solutions to important problems that face society.

An example of applied research is the research that led to the use of DNA to produce the drug insulin which is used in the treatment of the disease diabetes.

Applied research is usually funded and carried out by companies, the government, research institutes and the universities.

In the process of carrying out applied research important new basic information about nature may be uncovered. So the distinction between basic and applied research cannot be made too rigid.

Finally, research that is aimed at turning applied research discoveries into large-scale, marketable commercial concerns is developmental research.

When insulin produced through DNA technology was found to be efficacious in the treatment of diabetes, the research that resulted in its large-scale commercial production for sale by pharmaceutical companies is developmental research.

In Nigeria this quintessential continuum from basic to applied to developmental research is, as yet, virtually nonexistent.

The type of scientific research that a country should be doing can become a source of controversy. For instance, in some countries people are currently debating whether their government should be investing in basic or applied research.

The controversy is because, as I already mentioned, basic research whose chief purpose is to add to the pool of new knowledge, is very expensive and is funded primarily by the government and by all the citizens of a country through their taxes.

Private for-profit commercial companies use basic research information to produce marketable products for the financial benefit of only a few people in the society namely, their shareholders. This practice has become a source of controversy because many feel that it is unfair to the rest of society that funds basic research. I implore Nigeria's intelligentsia not to join this fascinating debate as we currently have more pressing problems at home.

Among Nigeria's numerous and persistent problems, one of the most pressing is her backwardness in science and technology. It is clear that most of the Nigerian public has continued to wallow in under-development today even though we achieved political independence over fifty years ago and are well endowed with human and natural resources. Nigeria's sorry predicament today is incontrovertible evidence that post-independence administrations have failed to utilize the acknowledged developmental tool of science and technology to improve the well-being of ALL Nigerians.

Since many people believe that the main reason for our underdevelopment relates in large part to a persistent impoverishment of science and technology in the country, a legitimate question is:

"Why has science and technology in Nigeria remained so impoverished"?

I submit that, *it is the failure of successive administrations since political independence to give adequate support to the education and training of our children in the sciences that mainly accounts for the impoverishment of science and technology in Nigeria.*

As it is incumbent on all of us to do whatever we can to combat this impoverishment let us first consider the manpower requirements for a successful science and technology enterprise in Nigeria.

I can identify at least three. Yours may be different; I'll explain mine as follows:

By far my most important manpower requirement for a successful science and technology enterprise in our country is the Nigerian President who is also the chief executive of the country.

I say the president because of his enormous capacity to set the national agenda and the far-reaching impact of his actions. For instance, if the president were to declare science and technology a national priority, governors as well as federal and state legislators will begin to give this enterprise the attention it deserves.

I also say the president because science and technology is a money intensive enterprise, and therefore requires a powerful and influential advocate in government since it is the government that allocates money for the enterprise. Furthermore the president is an important manpower requirement because he appoints the minister that oversees the country's efforts in science and technology.

However, to be successful as the most important manpower requirement the president has to be truly convinced that science and technology is a powerful enterprise for spurring development, and he also has to be willing to do what is necessary to upgrade this enterprise in Nigeria.

The second manpower requirement for a successful science and technology enterprise in Nigeria is, in my opinion, the president of the country's National Academy of Sciences. The importance of the National Academy of Sciences president lies in the fact that he is likely to have essential information on science and technology for development. He is, therefore, in a unique position to be able to provide the country's president, irrespective of his background, useful and unbiased information regarding the status of science and technology in the country.

My third manpower requirement for a successful science and technology is the people who actually carry out the work that becomes the nation's science and technology enterprise – the professors, scientists, engineers, laboratory directors, technologists and other technical and support staff.

Returning to the reasons for the impoverished state of science and technology in Nigeria, I reiterate my belief that ***the main reason is the lack of adequate support for the sciences in our secondary schools.***

In our system of education in Nigeria, instruction in science begins in secondary/high schools. Therefore if Nigeria's aim is to use science and technology as instruments of development she must provide our secondary schools with adequate support in science which is the seed-corn of technology. This is absolutely necessary for at least two reasons:

First, secondary schools are the ultimate source of indigenous scientists and technologists without whom science and technology cannot thrive in any country.

Second, if the sciences fail to attract young, bright secondary school students, these potential talents will not be available at the tertiary and university levels where the actual training in science and technology occurs.

I repeat, the sorry state of science in most secondary schools in Nigeria today is sufficient evidence of the lack of adequate support for the education and training of our children in the sciences.

If this situation is not changed for the better then the idea of using science and technology to enhance development in Nigeria will continue to remain an unrealizable wishful thinking.

Collaborating with advanced countries to use powerful foreign technologies like satellites to improve aspects of the standard of living of Nigerians as the country is currently attempting to do (*see chapter 2*) should be applauded but this does not, and will not make Nigeria as capable as she ought to be in science and technology.

Nigeria's effort to establish Sheda Science and Technology Complex near Abuja as the nation's premier research and development institute is admirable and commendable. But the institute's success will depend, to a very large extent, on the constant availability of talented indigenous scientists and technologists whose professional journeys would normally begin with Nigerian secondary school science.

Nigeria should grow her science from the bottom up, that is, from the secondary schools or better still from elementary schools, if she frankly hopes to be able to eventually use science and technology for the development of the country.

The world is now in a technological era, and Nigeria is an avid consumer of technology. Nigeria should not just be a big consumer; she should also position herself to be an originator of technological innovations or she will remain a dependent market that will keep enriching the advanced nations by relying on and buying their technologies.

With proper training in adequately equipped schools there is no reason why Nigerian children cannot mature to be real scientists and technologists that can, among other things, continually use their expertise to address local issues that are important in promoting the development of Nigeria.

Although science and technology is presently impoverished in Nigeria, this needs not always be the case and the situation can be reversed. After all, some of the present day modern societies that are now the bastions of science and prominent beneficiaries of its technological fallouts were not always in the forefront of the scientific research enterprise. And yet they have successfully incorporated science into their cultures and achieved their present high standard of living, in part by applying science and technology to meet their developmental and economic needs.

For Nigeria to uplift herself successfully from her current state of underdevelopment, she too must make a deliberate effort to incorporate science into her culture and apply indigenous technology to raise the standard of living of her people.

The ability to engage fruitfully in science and technology is not the birthright of any particular country or people, and yet too many Nigerians have the erroneous mentality that science and technology belongs to the advanced nations who are ordained to ration it out to developing countries as they see fit. Louis Pasteur (1822-1892), the father of Microbiology knew all along that this is not the case. Over two hundred years ago he said:

"Science knows no country, because knowledge belongs to humanity and is the torch which illuminates the world".

In other words, science thrives anywhere the condition is right.

Question: Professor, I was surprised that you did not indict corruption as being primarily responsible for the backwardness of science and technology in Nigeria today. Isn't it?

Professor: Thank you for your comment and question as they give me the opportunity to state publicly and unequivocally that I totally abhor the endemic corruption that has infested Nigeria and the damage it has done to retard development in our society. There is no question in my mind that corruption makes it difficult not just for science and technology to thrive but for any other developmental effort to have a chance of succeeding. So corruption per se in Nigeria warrants a special treatment, much more than this brief essay on science and technology can do.

However, the question that has continued to intrigue me and which I have tried to address here is:

If we continue to treat science in Nigeria's secondary schools the way we do today, is it realistic to expect that Nigeria will ever be in a position to successfully use science and technology, the acknowledged tools for social advancement, to solve our developmental problems?

My answer today is an unqualified "No".

Science and technology is carried out by highly-trained individuals, and you need a lot of such indigenous people in order for it to be successful in Nigeria. I say "No" in part because, currently we do not have enough indigenous educated manpower such as scientists, engineers, technologists and science teachers, etc. that can grow our science and technology locally. Neither do we have adequate training facilities in science for our citizens.

So I believe that Nigeria's development will remain stunted if we continue to neglect the sciences in our secondary schools as such an action will only continue to have a depressing influence on indigenous science and technology which is the acknowledged instrument for the development of countries all over the world.

I pray that both federal and state administrations in Nigeria will rise up to the obligation of initiating in our secondary schools the badly needed revolution in education in science which is the seed-corn of technology.

A role for individual Nigerians:

Every Nigerian should be concerned about the neglect of science in our high schools in view of the short- and long-term adverse consequences of this negligence to technology and development in Nigeria.

While it is the duty of our governments to lead in providing adequate support for science in our schools, all Nigerians that can afford it (and there are many of them) should feel a moral obligation to do their bit for science and technology in their country. By so doing they will not only be giving back to society but they will also be doing something worthwhile for future generations of Nigerians.

Specifically, individual Nigerians can help to improve the state of science in our schools by doing any or part of the following: build and donate science building(s) to a school, provide science equipment to a laboratory, donate up-to-date science textbooks or computers that are current to a school's library, sponsor the education of young, bright students who want to pursue science as a career or give scholarships to underwrite the training of prospective science teachers. If you have the resources you can even become a patron of science and provide financial livelihood to a talented young Nigerian so that he could devote all his efforts to science or technology.

If you don't have the financial resources to provide any of the above services, you can make time to talk to your political representatives about the necessity to give adequate support to science education in order to promote development in our beloved country.

The time is now for all concerned Nigerians to do something about the neglect of science in our secondary schools.

A prerequisite for a successful science and technology is a solid foundation in science which should begin in secondary or even elementary schools. The sorry state of science in Nigerian schools today is a national disgrace and an emergency. ***Go tell that to your legislators!***

The chapters that follow highlight the necessity for Nigeria to strive for indigenous competence in science and technology in order for her to use the latter to develop the country for the benefit of ALL Nigerians.

CHAPTER Two

First Nigerian satellites: the way forward

The following piece was written during a period of great turmoil in Nigeria, and after a lot of considerations the author decided to publish it. After all, there is really never a good time as it's always one thing after another.

One should be rightly puzzled that I would undertake to write about the country's satellite program at a period of national crises, with Nigeria having to deal with Boko Haram, infiltration of the administration by fundamentalist groups, unprecedented lack of personal security, untold hardships caused by the withdrawal of fuel subsidy in Nigeria, and an ongoing tragic and senseless abduction of hundreds of young Nigerian school girls whose only crime is the quest for education! And all these are occurring in the backdrop of endemic corruption and continuing lack of essential development of the country.

That one could write on subjects other than the above challenges indicates some sort of confidence that the government will listen to the protestations of well-meaning Nigerians that those who have undertaken to lead us should desist from political rhetoric and posturing, and rather lead by doing what is necessary to move the nation forward in the right direction.

That one can venture to address issues different from the above problems comes from the belief that our girls will come home safely. In fact, writing the essays in this book at this time is also recognition of the fact that the task of building the nation should not just be left to the government alone but that every Nigerian has a moral responsibility to contribute his or her efforts and talents to the task.

As the professor mauled over Nigeria's stubborn predicaments during his daily walk on a beautiful evening drenched in tropical sunset, an apparently happy and bubbly student approached him.

Professor: *Turning to the student, the professor said,* you seem awfully happy; you must have made a good grade in the last exam.

Student: Prof, it's the Nigerian satellites. I just heard that Nigeria successfully launched two satellites into orbit.

Professor: Really; new ones?

Student: Back in August.

Professor: But that's old news.

Student: Yes, it is. I wonder why I missed the news in August. I thought I was following Nigeria's space adventures pretty well. In fact, some years ago, I was all excited when reports that Nigerian engineers independently launched rockets from Lagos State University caused a great deal of controversy as to whether or not Nigeria really did accomplish such a feat.

Prof, what was launched in August, rockets or satellites, and this time, is the report true?

Professor: The report I read in August said that two new satellites, not rockets, were launched into orbit by Nigerian engineers. The Nigerian president even gave a news conference about it; so it should be true.

Am afraid I am on my way to a meeting that starts in a few minutes. Maybe we could continue this conversation some other time.

Student: Prof, I am sure many other students will be honored to take part in the discussion on satellites. If you don't mind, may I arrange for us to meet with you in a day or two?

Professor: That's okay by me. Just make it soon as I currently have a fairly flexible schedule.

The student then trotted towards one of the classroom blocks to mobilize some other students who in turn recruited their friends by means of the social media. The following day the large auditorium was filled with students from every level.

Apparently the news of the professor's informal talk had spread like wild fire throughout the campus. Because this meeting was held at the week-end there were no conflicts with regular lectures and so even some instructors and professors attended.

There were so many students that the informal discussion which had been scheduled in a small lecture hall had to be moved to a large auditorium.

Student #1 (Ms. Nneji): Thank you all for coming at such short notice. I am really embarrassed by this large turnout as this was not supposed to be a formal lecture by the professor who, at my request, graciously agreed to continue with an informal discussion of NIGERIAN SATELLITES that we started yesterday.

I have no prepared introduction, no nothing! So Prof, maybe you should just take over as I don't really know how to handle this.

The professor just smiled and moved to the podium.

Professor: Thank you, Ms. Nneji. I must confess that this is the best introduction I've had in my entire professional career.

I am just as embarrassed as you are by this large turnout but I'll like to reassure everyone here that this meeting will remain an informal discussion that was provoked by the recent launching of satellites by Nigeria.

As I said yesterday during my discussion with Nneji, I read recently that Nigeria launched two satellites and not rockets into orbit around our planet.

You can interrupt me whenever you have a question.

Student #1: Prof, what's the difference between a rocket and a satellite?

Professor: Both rockets and satellites are space crafts. They are fairly simple conceptually although complex in practice. I'll try to explain the basics of each simply and briefly, starting with rockets.

Although the principles of rocketry have been applied to many different contraptions over centuries, in the interest of time, I will limit my comments to the modern use of rockets in space travel today.

A rocket is a space craft or vessel that moves forward at a great speed when a high speed exhaust produced exclusively from a burning fuel or propellant in a rear or bottom compartment within the vessel produces a high speed thrust that moves the vessel forward in the opposite direction. The burning propellant is visible to the naked eye as flame and/or massive smoke in the rear of the rocket (Figure 1).

Figure 1: A US rocket being launched into space (Courtesy of the National Aeronautics and Space Administration-NASA- of the USA).

In order words when the rocket is in a vertical position, the high speed exhaust produces a powerful force downwards and the rocket reacts by moving upwards with an equal force as is predicted by Newton's third law of motion which states that for every action there is an opposite and equal reaction (Figure 2).

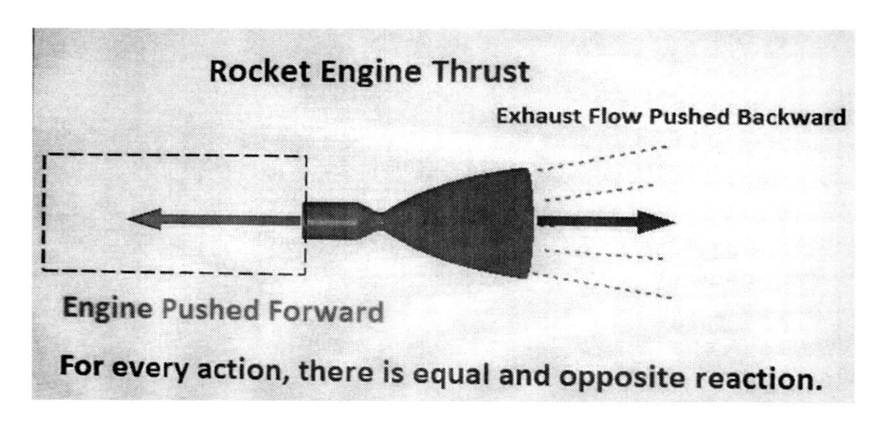

Figure 2. Diagram illustrating the genesis of the forward movement of rockets (Courtesy of NASA).

However, some have argued that the primary reason for the forward movement of the rocket is that the burning fuel in its rear-end creates a huge but unbalanced pressure inside the vessel and that it is the tremendous unbalanced pressure against the upper inner surface of the chamber that propels the rocket forward at very high speed.

Unlike a rocket, a **satellite** is a body that is in orbit and moves around a celestial body in space. A modern rocket is usually designed to serve as a vessel that carries a satellite to its required orbit in space. Advances in the speed and altitude that rockets can now attain have made it possible for satellites to travel through astronomical distances to explore space.

A satellite is typically put into its orbit in two to five steps with the help of a rocket system called launch vehicle (Figure 3) which carries the satellite into space.

Figure 3. A launch vehicle (on the left) for a satellite housed in its uppermost compartment (Courtesy of Boeing Company).

The vehicle is made up of several segments, also called stages by some (Figure 4).

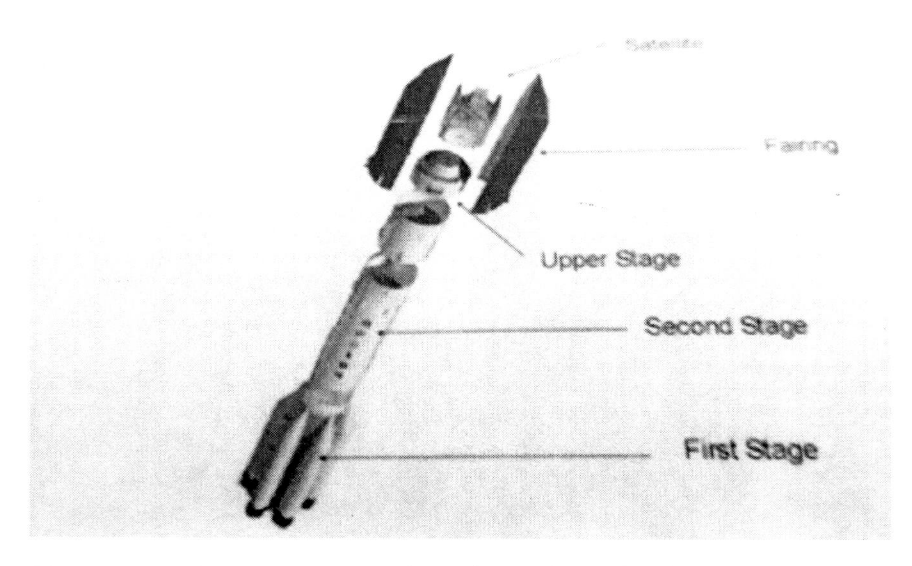

Figure 4. Components of a launch vehicle (Courtesy of International Launch Services).

The first segment (stage) of a launch vehicle consists of rockets and a fuel tank (Figure 3 & 4). When the rockets of the first stage are ignited, the launch vehicle housing a satellite in its top compartment is lifted from the ground towards the sky. At lift-off, the launch vehicle is very heavy so the rockets of the first stage are multiple and have to be very powerful. After the rockets' fuel is exhausted the first stage is no longer needed and it is jettisoned from the launch vehicle. This ends the first step of the launching.

The second segment (stage) of the launch vehicle also has rockets and fuel tanks, and the second step in the launching begins when these rockets are fired immediately after the first segment is discarded. The launch vehicle is now carried into space, and when the second stage rockets' fuel is exhausted the second segment again separates from the lunch vehicle and burns in the earth's atmosphere.

The uppermost end of the launch vehicle called the "upper stage" is connected to the satellite which is housed in a protective metal shield called "fairing" (Figure 3 & 4). Once the satellite is in space above the earth's atmosphere the fairing splits open. The rockets of the upper stage fire and put the satellite in the desired orbit.

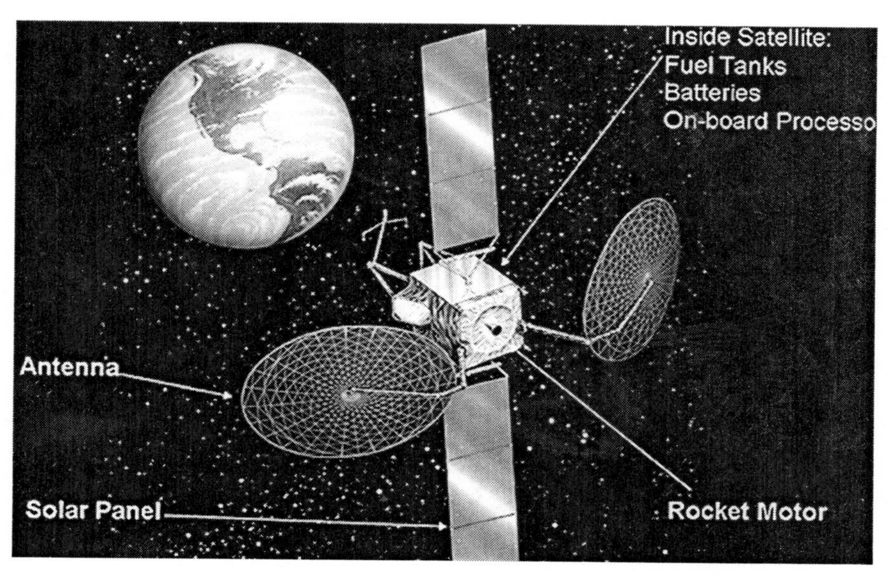

Figure 5. A communications satellite in orbit with its extended solar panels and antennas (Courtesy of Boeing Company).

The satellite then unfurls its solar panels and communication antennas which had hitherto been stored away, and the satellite's mission is begun as it moves round the earth (Figure 5).

Once in orbit, the earth's gravitational pull on the satellite is balanced by its inertia, so the satellite does not need external energy from propellant/fuel to move around the earth in its orbit.

This does not mean that satellites do not utilize energy or power. A satellite needs energy to perform the various duties that are assigned to it. Some satellites rely on fuel cells or batteries or even nuclear sources for their energy. Others can generate electrical energy using solar panels and solar cells and the sun as battery charger.

Student #2: *Fascinating!* Prof, do satellites have other significant characteristics?

Professor: I think it is worth emphasizing that unlike a rocket, a satellite usually moves around another object. For instance, the earth is a satellite since it moves around the sun, and the moon is also a satellite since it moves around the earth. The earth and the moon are sometimes called natural satellites to distinguish them from artificial satellites which are man-made and launched in orbit by human beings.

The first artificial satellite is Sputnik 1 which the Soviet Union launched into orbit around our planet on October 4, 1957. Although the Sputnik lasted only about three months before it dropped out of its orbit and burned in the earth's atmosphere, its successful launching was a remarkable achievement as it ushered in the current space age.

Since 1957 thousands of man-made satellites and space "shuttles" have been put in orbit around the earth by various countries.

Only a few of these are operational while the rest are essentially junks polluting space. These space junks consist mainly of unusable satellites, satellite fragments and rocket fragments.

Students #3: What are shuttles?

Professor: Space shuttles are not satellites. The official name the US gave the space shuttle is Space Transportation System (STS). The STS program lasted for thirty years, from 1981 to 2011 when the last shuttle was grounded. The space shuttle was conceived as a space craft that could carry humans to and from space.

The US planned to make the space shuttle at least a partially reusable space craft because space crafts are expensive to build. The idea was that with the shuttle, the same space craft could be refurbished and reused after its return to earth from its mission in space thereby reducing the cost of building a new one from scratch.

The US built a total of five shuttles. One exploded in 1986 during take-off and another broke up during re-entry from space in 2003. In both of these unfortunate accidents, all the crew members were killed. Our prayers go to the dead astronauts and their families for their enormous contributions to science and humanity.

The shuttle consists of three main parts: the orbiter, a large fuel tank (external tank) and two solid rocket boosters (Figure 6).

Figure 6. Components of US space shuttle (Courtesy of NASA).

The two solid rocket boosters and the shuttle's main engines at the base of the orbiter were pre-programmed to ignite and initiate take-off, and the gasses they burned at their rear ends provided the thrust for the space shuttle's lift-off (Figure 7).

Figure 7. A US space shuttle at liftoff (Courtesy of NASA).

The solid rocket boosters were jettisoned about two minutes into ascent, and were picked up from the ocean to be refurbished for a future flight. The shuttle's flight to its orbit was continued with energy provided by the orbiter's main engines which were fueled by the external tank.

The orbiter was the part of the shuttle that carried the crew and the shuttle's payload. Because it was the only part of the shuttle that entered orbit, some regard only the orbiter as the shuttle. This is technically not correct because the official name of the shuttle is space transportation system which consists of the orbiter, two solid rocket boosters and an external fuel tank.

Like a satellite, a space shuttle was launched into orbit by conventional rocketry. Part of the shuttle, the orbiter stayed in space in its orbit like a satellite. Since the orbiter like satellites did not need energy to move around in its orbit the external fuel tank was jettisoned and it burnt in the earth's atmosphere. So the external fuel tank was the only part of the shuttle that was not reused in future flights. The orbiter's mission in space usually lasted for one to two weeks after which it exited orbit and started its return journey back to earth.

The descent from space involved complex maneuvers that were aimed at transiting the orbiter from a space craft flying at about 17,000 mph to an airplane flying at about 250 mph when it entered the earth's atmosphere. As we saw earlier, with the jettison of the external fuel tank there was no longer any fuel available to the orbiter; so the latter made the return journey to earth as an unpowered high-tech glider.

One astronaut described the orbiter as basically falling rather than flying during its return from space. However this was a complex, well-coordinated falling that was pre-programmed into the orbiter during its design.

Most of the time the return journey was navigated by autopilot, but the shuttle's commander manually piloted the orbiter during the final stages of its descent to earth. On landing, a parachute was deployed at the rear of the orbiter to help it cruise to a stop (Figure 8).

Figure 8. The shuttle's orbiter landing as an airplane after completing its mission in space (Courtesy of NASA).

America's hope of building, a partially reusable space craft was fulfilled because the orbiter, as were the jettisoned solid rocket boosters, were refurbished and reused for another space flight.

You have to agree that the space shuttle is a technological marvel.

Student #4: Prof, does the space shuttle have anything to do with the International Space Station?

Professor: Although satellites perform many important functions, they do not carry humans and, as a result their usefulness is restricted to what they are programmed to do before they are sent to space. Industrialized nations realized early that in order to conquer space, what they really needed is a space station where they could have manned laboratories to perform frontier research in many fields including space weather, human biology, medicine, physics, astronomy and meteorology, etc.

Furthermore, if space is to be successfully colonized as industrialized nations hope to, they need to know the long-term effects of space exposure and weightlessness on the human body.

As a result tremendous expense and expertise are required in order to understand and conquer space. Consequently, the national space agencies of America, Russia, Japan and Canada combined to form what eventually became a 16-nation International Space Station (ISS) in order to share the burden that would have fallen on a single nation.

The ISS is conceived as a complex, well-equipped, futuristic, orbiting laboratory with living and experimental quarters for up to seven astronauts at a time. At completion, the ISS will be larger than a five-bedroom house; will weigh almost one million pounds or the equivalent of more than 320 automobiles, and it will measure 357 feet end to end or slightly larger than full-sized soccer field.

The space shuttles were initially used to place astronauts, satellites and equipment like the Hubble telescope in space. Because the ISS is so huge and complex with vital tasks ahead, an important function of the shuttle became tied to the ferrying of humans and materials (Figure 9) to complete the building and equipping of the ISS.

Figure 9. An American space shuttle ferrying astronauts and materials to the International Space Station (Courtesy of NASA).

Another important reason for the joint global space effort in the ISS project had to do with national security. Space exploration began during the cold war when nations were obsessed with military might and superiority over others. Industrialized nations were suspicious of one another and feared that if any one nation built an ISS, she could use it militarily to dominate the others. They reasoned that a joint effort would make this less likely.

Student #5: Now that America has retired her last space shuttle, does it mean that the ISS is completely assembled?

If not, without a shuttle, how will American astronauts and the international community get to the ISS in the future?

Professor: Good question. The ISS is currently habitable and there has always been an astronaut working there for the past fifteen years. Its fabrication is not yet complete and humans and cargo continue to be ferried there (Figure 9). Only the Russians now have the space craft that can take astronauts to the ISS.

Although this is disconcerting to some, the Americans and the international community now depend on Russia to ferry humans and cargo to the ISS that was initiated by America.

So with their space shuttles now grounded, Americans are in the era of going to space by hitching rides with Russian astronauts. In fact since July 2011, American and Russian astronauts have been flying jointly in Russian space crafts to the ISS.

The American administration believes that eventually suborbital tourism by private commercial companies which are now evolving in the United States will become so common that for a fee they would transport American astronauts and cargo to space independent of the Russians. This will free American astronauts to continue the exploration of deep space which is what they are really interested in doing.

Student #6: Your last statement partly addressed my question which is, with the retirement of the shuttle program, what will become of American space program and the National Aeronautics and Space Administration of America (NASA)?

Professor: That's a question that many Americans interested in space and its implications have also been asking. The American president dealt with the issues raised by the question when he visited Kennedy Space Center, Florida three months before the space shuttle made its final flight.

President Obama assured Americans that the exploration of space by NASA will continue when the shuttle program ends in a couple of months. He said that future space program was necessary to continue to increase our knowledge about the world and its climate in order to protect the environment for future generations. The shuttle program, he assured, was stopped in part to enable America to prepare for the next stage of space exploration into deep space to the moon and as far as to Mars.

In this new phase of the space program he continued, destination would no longer be the main object. But the new space program would now include doing what is necessary to enable people to live and work safely beyond the earth for as long as possible.

The president challenged NASA to quickly make the breakthroughs in advanced technologies that will facilitate the new phase of space travel and make eventual travel to Mars safe and affordable. Since he also promised to give them the necessary resources, the American space program and NASA live on.

Why am I telling you all these?

I am because I think it is important for you as students and Nigerian youths to know what is happening around you, and to realize that the space story is one of remarkable achievements that resulted from the undertaking of difficult but worthwhile challenges as those around you today. I am because I know that you too have it in you to be able to perform technological marvels in your environment.

How do I know that?

I know because whenever your counterparts overseas are plugged into programs of advanced technology,or in fact any important program, they too perform admirably and sometimes beyond expectations. I know because knowledge does not belong to any particular nation or people. It belongs to all humanity and you are part of humanity. I know because it is the quest for knowledge that caused this large auditorium to be filled voluntarily at a moment's notice by eager students anxious to learn more about the achievements of their fellow countrymen in satellite technology.

And I will agree if you say that you cannot perform technological marvels at this time in your country because of the current inadequacy of science and technology in your environment.

Nevertheless, you should know that the industrialized world saw space as an unexplored frontier with potentials they could hardly fully foresee, and yet they embarked on doing what they could to conquer it. They did this by mobilizing their country's best resources in academia, politics and society. And those entrusted with the responsibility of spending the millions and sometimes billions of dollars of public money required to execute the space program used the money for what it was intended.

Everybody, including you, has some talent. If you don't know yet what yours is, this is when to find out, and please plan on using it.

You as students who will inevitably become the future leaders in a country replete with unnecessary human sufferings, always be aware that real progress requires hard work. It also requires dedication and the giving of your best for the benefit of your country and her citizens even if it means making the ultimate sacrifice like the astronauts that died in the shuttles' fantastic ventures into space.

The people in the forefront of the space program today are industrialized nations that have met their basic needs of food, water and shelter and therefore can afford to explore space. Nigeria has not met her basic needs, and it seems to me that we as a nation have so much to do here in our environment to improve the lot of our people.

We have all the basic ingredients for an exciting and successful future – people, natural resources, and a nation full of great challenges. We are still at a stage in our development when we do not need to be geniuses to know what we need such as reducing the high level of illiteracy in our midst, building adequate roads, providing reliable electricity and an environment that will facilitate the production of food, water, good sanitation, shelter, education and jobs.

A frightening development in any society such as ours is to continue to have a large pool of idle, young people who do not have anything positive or challenging to occupy their minds especially when they themselves know that there must be something worthwhile they could be doing. Events in our country are daily reminders that a mind that is not meaningfully occupied has a way of drifting towards something that is unsavory.

So I implore you again to seek out what your talent is and begin to utilize it for the benefit of your country and humanity.

Although you may not realize it now, please believe me when I tell you that life, especially productive human life, is unfortunately very short.

Sorry, I digress!

Are there any questions about space shuttles?

Student #7: Prof, what did America do with the retired shuttles?

Professor: America had three historic surviving space shuttles, specifically orbiters, and they were retired when the space shuttle program ended in July 2011.

At their retirement, twenty one institutions across America vied fiercely for the opportunity to be the shuttles' permanent homes where they would be displayed in museums for the public to view and enjoy.

The competition was fierce because the shuttles are icons and any community housing a shuttle would likely enjoy a tourism boom.

After a lot of considerations NASA awarded the three historic shuttles, Atlantis, Endeavor and Discovery to museums in Florida, California and Virginia, respectively.

The fourth shuttle, Enterprise was a prototype test shuttle that was used for early glide tests but was never sent into orbit. NASA awarded it to a space museum in New York.

Those cities that were not selected as home for any of the retired shuttles were, naturally, deeply disappointed. Some even impugned that NASA's decision not to award the shuttle to some cities was politically tainted.

A distraught Texas legislator said that the oversight of his state "smacks of a political gesture in an Agency that has always served above politics". Another bemoaned that NASA's decision not to choose Texas as a permanent home for one of the space shuttles slighted the numerous important contributions that Johnson's Space Center in Texas made to the space program.

The official position of President Obama's administration is that the locations chosen for the retired space shuttles will "ensure that the greatest number of Americans will have a chance to see these national treasures and learn more about their significant contributions to national space exploration history".

The shuttles did not all fly to their retirement destinations. Atlantis was towed just a short distance in the Kennedy Space Center, Florida to its final location in the same Center. The rest travelled piggyback on Boeing 747 to their various destinations (Figure 10).

Figure 10. A space shuttle riding piggyback on Boeing 747 to its final destination (Courtesy of NASA).

Are there any questions about Nigerian satellites which are the catalyst for this discussion?

Student #1: Prof, could you say something about the satellites that Nigeria launched into space recently and why some people were so excited about the launching?

Professor: I'll just paraphrase what was attributed to the President of Nigeria at a news conference following the launching. According to President Jonathan, Nigeria launched into orbit in August 2011 two satellites named NigeriaSat-2 and NigeriaSat-X. Both satellites were designed and built by

Nigerian engineers/scientists in Britain in conjunction with British scientists and they were launched into orbit in southern Russia aboard a Russian rocket. The progress of the satellites is being monitored from control stations in Guildford, United Kingdom and in Abuja, Nigeria. NigeriaSat-X was designed and built exclusively by a team of Nigerian engineers and scientists as an experimental satellite.

The Nigerian president was exuberant about the successful launching of the satellites and congratulated the Nigerians who made it possible. He said the satellites would boost Nigeria's capacity to deliver Internet services, provide information that will be useful in: weather forecasting, watching natural disasters like floods, urban planning and the monitoring of crops in farms.

Personally I found it exciting to see the first very high resolution picture produced by a Nigerian satellite although it would have been even more thrilling if that first picture was a Nigerian landmark and not an airport at Salt Lake City in the USA.

Many others were rightly moved by the successful Nigerian satellite launch because of the many useful tasks satellites perform and the numerous ways that a country can use satellites to improve the lots of her people. The president already alluded to some of these functions of satellites which are now commonly used to improve communication both nationally and internationally.

Satellites can also be used to aid in searching and rescuing individuals in remote and dangerous places, and in seeking distressed ships and aircrafts as in the recent Malaysian aircraft tragedy of March 8, 2014.

While a satellite does not give you an ideal weather, satellite technology can assist you in telling and predicting the weather in any part of the world as is evident on television every day.

Satellites have unlimited capacity to view the entire globe and take pictures of any nook and cranny of the earth. So satellites can provide information on the earth's resources, and uncover environmental issues such as landslides, mudslides, gullies, oil pollution, or just the topography of the land.

Satellites assist humans to make predictions of disasters such as dangerous storms, flooding and tsunamis so that corrective measures can be taken even before these occur. They can also provide a good idea of population densities by determining the fraction of a region that is inhabited. Satellite-TVs can be used for educational purposes in regular classrooms as well as in obscure parts of any country especially those without formal educational resources.

One of NASA's biggest achievements in this space age is the launching of the Hubble telescope in space with the aid of a satellite. Astronomers use the Hubble telescope to continue the study of the universe by making observations and producing crystal clear images that are impossible to make from the ground because of unavoidable interferences from the earth's atmosphere.

Satellites have become indispensable for military purposes and are used for intelligence gathering and reconnaissance. Although their military role is understandably guarded in secrecy, it does not seem to be waning, but luckily as you have seen above it's now only one of the uses of satellites.

In short, the functions of satellites are only limited by the degree of sophistication of one's imagination, as there are so many important facets of human existence that can benefit from a judicious use of satellites.

One of the lessons that students should learn from this discussion on satellites is that a technological advancement is not necessarily an end but rather can become a tool for addressing old problems.

For instance, in the nineteen fifties, because of the perceived usefulness of the then nascent telephone, advanced nations struggled with how telephone signals could be relayed across the Atlantic Ocean to Europe. It was near impossible to lay cables or string overhead wires across the vast ocean to carry telephone signals as is usual for conventional telephones across land masses.

But after the Russians launched the first artificial satellite into orbit around the earth in 1957, industrialized nations quickly applied the new technological development to address their age-old problem of transatlantic telephone communication. The Americans began to experiment on routing telephone signals first to an orbiting satellite which then relayed the signal back to the earth to telephone masts across the Atlantic Ocean.

By 1960, the American telephone giant, AT&T had applied to the US government for permission to launch the first experimental communications satellite. Today the entire world is linked by satellites not only for telephone but also for television services.

This clearly illustrates the use of new technology to solve an old problem, which is a staple in science and technology.

Another example of the use of new technology to solve an old problem is the development of navigation satellites. Industrialized nations had for a long time struggled with the fact that the navigation of ships was compounded by the fact that there was no easy way to determine the exact position of vessels that were in the ocean out of sight of land.

However, it occurred to US's John Hopkins University scientists monitoring the 1957 Russian Sputnik that satellites could be used to decipher the position of a vehicle anywhere at any time. Their work in this area led to the development of a technology that is now used to locate exactly where ships are at sea or in the ocean.

The above navigational technology also led to the development of global positioning system (GPS) which is now installed in most late model cars in industrialized nations. GPS utilizes satellite technology to pilot a driver to any location known or unknown. This is accomplished simply by typing in one's destination in a GPS receiver installed in the vehicle and pressing start, and the instrument directs you to your location by means of a map, verbally or by both methods. Now there are even portable GPS devices and cell phones with built-in GPS for those that do not have one that is factory-installed in their vehicles.

I think it is easy to see that the president and the interested public were excited about the successful launching of Nigeria's satellites because of the many potential applications that satellites offer.

Unfortunately, a satellite is very expensive. Since a single satellite cannot accomplish all the different tasks that satellites can perform, a satellite is usually built with specific function(s) in mind. This contributes to make the acquisition of satellites an expensive undertaking. The cost of a satellite varies depending on the size, the materials used to build it, its payload or what it's carrying and the job it is supposed to accomplish. However, no matter the type, satellites cost hundreds of millions of dollars to design, construct, launch, monitor and maintain after they are launched. Commercial satellites are even more expensive and their cost can run into billions of dollars.

Student #8: Prof, from all you've said it sounds like every country will profit from the use of satellites. But at the same time it seems that not every country can afford them because they are so expensive.

My question is, is it possible for a country that cannot afford to build and launch its own satellite to get the information it requires simply by using other peoples' satellite for a fee?

Professor: That's a good question and one that many people have wrestled with. One journalist expressed essentially the same sentiment a little differently. The journalist noted that many developing countries are building similar expensive environmental satellites that have the capability of taking pictures of any part of the earth. And he wondered if it was prudent for countries to engage in building expensive satellites when they have more urgent problems like feeding their people.

The journalist then lamented that the problem becomes more

worrisome when one considers that free satellite data are available from the United States and Europe.

Let me first disabuse you of the notion of free data. If something is said to be free, you should examine the circumstances very carefully to see if it is truly free and if it actually meets your needs.

Secondly, you should reflect on the fact that the hidden costs of the so called "free" satellite data may become so exorbitant that funding the building of your own satellite may turn out in the long run to be cheaper especially when you consider that the source of the apparently free data could be withdrawn at a moment's notice.

Be aware that if you focus only on cost you'll miss an important point because developing countries will then always remain consumers waiting for the technological innovations of the advanced nations.

Developing countries, by having their own satellites, will be putting themselves in a position where they could one day become the experts in addressing problems that are peculiar to their region through the use of satellite technology.

Furthermore, it has been suggested that developing countries may be trying to build their own satellites in order to help promote their national self-confidence and to develop their countries' space scientists and engineers and thereby generate indigenous role models for their children. The importance of this cannot be measured in dollars and cents.

Student #9: Prof, I noticed that the Nigerian satellites were actually built in Britain and launched in Russia.

Can these satellites be really characterized as the product of indigenous effort by Nigerian scientists and engineers?

Professor: In answer to your question, you might like to know that it is not unusual to have an advanced country make and launch the satellites of other countries especially those of developing countries in the beginning of their space adventure.

I hope you will agree that it is not necessary to expect developing countries to go through the trial and error that advanced nations had to go through to develop satellites.

Yes, Nigeria's satellites were built in Britain and launched into orbit from a launch pad in Yansy, Russia. The lamentation by some that Nigeria's satellite effort is therefore foreign-based is misplaced because this is exactly how it should be at this stage of Nigeria's development.

In fact, the launching of the world's numerous satellites is done by only ten countries. On the day that Nigeria's satellites were launched, Russia also launched satellites for six other countries.

Nonetheless, every country should consider several issues before embarking on building and launching satellites. One of these issues is, of course, cost. We already saw that a satellite is very expensive and could cost millions of dollars if everything went well.

The price tag could run into billions if something went wrong like: damage to the vessel, the payload does not function properly or something goes wrong with the launching.

The immense monetary investment required by a satellite becomes a source of worry if the developing country acquiring it is one in which the rulers seem to be more interested in their personal welfare than in the welfare of the nation, and so do not apply public money to what it is designated for.

All along, we've been buying the products of technology: electricity, cars, trains, airplanes, medical technology, communications technology, etc., and now space technology. Initially it's okay to buy technology if you can afford to do so.

But you must also position yourself to be a producer of technological innovation. That's why Nigeria must invest in our schools and in the training of our children today in science and technology.

It is not enough for our engineers to be able to operate other people's technology. Anyone with the appropriate background can be trained to do that. What is more important is to create an environment locally where our own people can make original contributions to the pool of world science and technology. Industrialized nations have known and practiced this for centuries.

What happened after the Russians launched the Sputnik in orbit in 1957 was not just happenstance. The Sputnik's launching motivated progressive governments around the world to promote science in all their schools. In the US, science homework increased in schools, and children's interest in science and technology was promoted and encouraged in all the media because industrialized nations could see the potentials inherent in putting a satellite in orbit.

They also know that for today's children to become tomorrow's always-needed scientists and technologists their interest in science and technology has to be captured early.

At the beginning of the space age some countries even recruited engineers from other countries to teach them how to build better engines for space travel and at the appropriate time, they sent them back home. In Nigeria we too must understand that self-reliance especially in science and technology is a must and that we should do whatever it takes to promote these enterprises locally if we really intend to foster development in our country.

Not too long ago, the modern computer was born. Americanlegislators quickly realized its potential and their governments immediately mandated appropriate electrical wiring of all elementary schools so that computers could become accessible to all school children, and eventually to every American.

They did not let personal ownership of computers become a status symbol. Rather they realized that it is in the long term interest of their nation that everyone is computer-literate so that they could participate fully in the impending computer based revolution of their economy.

Those in positions to take similar steps in Nigeria have a moral obligation to do so forthwith. I am by no means implying that a developing country like Nigeria should compete with the industrialized world in science and technology; far from it in view of our current condition.

What I am suggesting is that we position ourselves in a state to be able to intelligently and judiciously utilize and apply advances in world technology to address our local needs.

Every Nigerian interested in the use of technology to advance our development should ask him- or her-self, "how is science in my elementary school, high school, tertiary institutions and universities"? If your answer is "not good", then you are obligated to do all you can to reverse that answer because we are now in a technological world, and science is the seed-corn of technology which is indispensable for our national development.

I believe that a leader firmly convinced of the power of the use of science and technology to promote development will give these enterprises the magnitude of the support they deserve.

Since you the students of today will surely become the leaders of the future, I urge you to survey the standard of living of the peoples of the world and convince yourself that science and technology is really a proven stimulator of development around the globe. As students, irrespective of your circumstances now, consider yourself privileged, and pledge that you will do whatever you can whenever possible to promote science and technology in Nigeria both now and in the future.

First Nigerian nuclear power plant

A STUDENT KNOCKS on the Professor's office door.

Professor: Come in.

Student: Good afternoon, Prof. Am sorry I have no appointment; but can I see you for a minute?

Professor: Of course you can. You know I have an open-door policy. Come right in and sit down. What's on your mind?

Student: I've been mulling over a problem for some time, and I've concluded that I'll profit from discussing it with someone more knowledgeable than I am. It's not really a problem as such. But thinking about it worries me and..................

Professor: Just go to the point.

Student: Recently, I read that our country is planning to build a nuclear power plant to produce electricity. Is that so?

Professor: I've heard that too. I understand the Federal Government of Nigeria has contracted with the Russian Federation to build it.

Why is this problem for you?

Student: I've read so many nightmarish accounts of nuclear power plant accidents recently, so I resolved to learn more about nuclear plants.

Professor: Actually, the decision to build a nuclear power plant is a political one, so I hope the president will educate the Nigerian public on the pros and cons of nuclear power and why he thinks Nigeria needs one now. But as a citizen I commend you for striving to be better informed about such an important potential undertaking by our government.

The professor then went on to explain as follows:

As you know Nigeria is plagued by absolute and relative lack of electricity depending on where you live. Power outages and "low current" are the norm rather than the exceptions even in affluent neighborhoods. Many institutions can hardly operate fully in Nigeria in part because of erratic electricity supply. Companies that do establish in Nigeria do so at their own risk because many as well as some private individuals have to resort to private generators for their electricity needs. Schools, hospitals, clinics, banks and many essential services are obliged to operate with suboptimal levels of electricity or private generators. At night, especially in the villages, the country is covered by a blanket of darkness from as early as 6:00 pm in some areas, and many people make do with lanterns and kerosene stoves.

You're already familiar with all these, so let's go back to power plants. Yes, nuclear power plants can be used to generate energy and many developed countries use it to supplement their electricity energy needs.

Fossil fuel (from coal, oil and natural gas, the traditional sources of energy for electricity), however contributes about 80% of global greenhouse carbon dioxide. As a result fossil fuel has been implicated in the undesirable warming of our planet, which is believed to cause the extremes of weather the world has been experiencing recently.

But while the splitting of atoms like uranium or uranium fission in nuclear power plants also causes the production of energy, in contrast to fossil fuel burning the process does not create undesirable carbon dioxide as byproduct as well. So nuclear power plants can produce clean environmentally friendly energy for generating electricity.

A country's electricity comes from different energy sources. Furthermore the amount contributed by each source varies from country to country and from time to time. In 2012 the US generated 68% of her energy from fossil fuel (coal, oil, and natural gas), 19% from nuclear energy and 7% from hydropower and the rest from other sources.

In the UK in 2011, petroleum oil provided 42% of her total energy production, natural gas provided 33% and coal 8%; nuclear, wind and hydro energy provided 13% and bio energy and wastes accounted for the rest.

In France, nuclear power is the main source of energy and provides almost 80% of her electricity needs, the highest in the world. Following the Chernobyl nuclear accident in 1986, anti-nuclear protesters warned France to shut down aging nuclear plants, and after the Fukushima nuclear disaster of 2011, the French public became increasingly concerned about the risks associated with nuclear power plants.

Italy abandoned nuclear energy production after the Chernobyl nuclear disaster in 1986, and she has no functioning nuclear power plants now. In 2011, 47% of her electricity was from gas, 17% from coal, 16% from hydro power and 6% from oil. Italy recently announce the building of the world's first hydrogen power plant. This is commendable since hydrogen is a clean energy carrier; however, the age-old problem of how to produce hydrogen for the plant without the use of atmospheric polluting fossil fuels persists.

America and many advanced nations are currently trying to markedly reduce the degree of their reliance on energy that comes from coal, oil and gas in favor of environmentally friendly sources like wind energy, solar energy, geothermal power plants, etc. When they succeed, countries whose economies are currently based essentially on oil production cannot say that they didn't see it coming. Today, such countries should in fact consider themselves blessed because while advanced nations are looking for alternative energy sources they can afford to devote some of their own efforts to diversifying their economies, at the least, in order to minimize the impact of a diminution in the world's oil needs that will come in the future.

Student: Why is nuclear energy such an undesirable source of electricity?

Professor: That's a good question. But let's first briefly and simply review how energy is derived from nuclear power plants which Nigeria intends to build.

The chemical element uranium is very energy dense as the amount of free energy it contains is millions of times the amount in a similar mass of chemical fuel such as gasoline. Consequently uranium is utilized by most nuclear power plants as a source of energy.

Within a nuclear reactor in a power plant the uranium atom is split into two by a chemical reaction and the process results in the release of energy in the form of large amounts of radiation and heat. The heat is used to generate steam which in turn is used to drive the turbines that produce electricity as in fossil fuel plants.

Student: So another difference between nuclear power plants and fossil fuel plants is that nuclear power plants in addition to producing energy in the form of heat also produce large amounts of radiation.

Professor: You are absolutely right, and that is an important difference. It is this undesirable by product, radiation that has made nuclear power such a controversial method of producing electricity. In fact the disadvantages of nuclear power plants relate to the deleterious consequences of radiation on humans and animals.

The opponents of deriving electricity from nuclear plants are fearful that nuclear power plant accidents will result in catastrophic effects because of enormous release of radiation into the atmosphere where it can cause genetic mutations and cancer in humans and animals including fish. Young growing children are particularly vulnerable to the deleterious effects of nuclear accidents.

The radiation can be carried by wind to faraway places so that the deleterious consequences of nuclear accidents are not limited to nearby areas. These fears were substantiated by a serious nuclear power accident in Chernobyl (Ukraine) in 1986 which we've already alluded to. In this nuclear plant disaster radioactive fallout was not limited to Chernobyl area but radioactive materials also contaminated large areas of the then Soviet Union, the Scandinavian countries, the Netherlands, parts of Europe including Germany, northern Italy, eastern France, Greece, the United Kingdom, etc.

Thousands of people were evacuated from the most contaminated areas leaving behind several "ghost" towns. One gets an eerie feeling while viewing these ghost towns in you-tubes still posted in the Internet.

Reports of the total number of fatalities due to the Chernobyl nuclear accident vary. In one account soon after the disaster 47 people died, some from acute radiation poisoning. In another report death from cancer and genetic defects was estimated to be as many as 28,000 to 100,000 within 50 years of the disaster. While these numbers may be on the high side, available evidence indicales that many people have died as a result of the Chernobyl nuclear accident.

The answer to your last question is now obvious. Today nuclear energy is not particularly widely used for electricity generation because many people are fearful that radiation from a nuclear accident will result in horrible genetic mutations and cancers in humans and animals as well as contaminating the environment especially the seas and farms and the food they produce.

Student: Now it makes sense that the Japanese were demonstrating against nuclear energy during their nuclear plant accident.

Professor: Yes, as you know in March, 2011 an earthquake hit Japan and set off a tsunami that damaged nuclear power plants in Fukushima, Japan and this caused the spread of radiation into the atmosphere, the soil and the Pacific Ocean.

While awaiting the full report of the effects of this nuclear disaster, its consequences were already being felt around the world. For instance, Italy immediately extended the moratorium it imposed on nuclear power plant construction after the Chernobyl nuclear accident in 1986. And the German Chancellor, Angela Merkel, announced that her government would phase out all of the country's 17 nuclear plants by the year 2022.

Although the chancellor did not expect in Germany the type of natural disaster that hit Japan in 2011, the German chancellor said she was dismayed by the helplessness of a technologically advanced nation like Japan after the Fukushima nuclear disaster. She decided to abandon nuclear energy in order not to inflict on her people the type of humanitarian disaster caused by the Fukushima nuclear accident.

I shudder to think what the Nigeria of today would do if faced with a nuclear power plant accident in which massive amounts of radiation are released into the atmosphere.

Student: Why is Nigeria proceeding with the building of a nuclear power plant?

Professor: Remember, my dear student, that in the beginning I told you that the decision to build a nuclear power plant is a political one. So only the Nigerian president can truly answer your question of why Nigeria is proceeding with the building of a nuclear power plant at this time.

It is a truism that when nuclear disasters occur most governments step back and reexamine their own nuclear ambitions because of the possibility of disastrous exposure of their people, animals and farmlands to radiation.

Nigeria's nuclear adventure and the nuclear disasters that occurred in Fukushima and elsewhere remind me of the story of the sailor whose boat was sinking and when another boat sailed by to rescue him he said, "No, God will save me". As his boat kept sinking a second boat came to the sailor's rescue and he still refused to join him because he believed that God would save him. Then a third boat came by and pleaded with him to join him, and he again said, "No, God will save me". Soon after, the sailor's boat sank and he drowned and died. In heaven the sailor walked directly to God and demanded to know why God abandoned him. *No way, God said, I sent you three boats and each time you refused their pleas to join them to safety.*

I believe that before embarking on the use of nuclear power to generate electricity, a government should first exhaust the use of safer traditional methods. It should also have in place essential public infrastructure such as adequate roads, efficient health-care services, clean water and schools that produce indigenous scientific and technical manpower.

In addition it should be a common practice that before embarking on building and operating a nuclear power plant, a country should have a ready pool of indigenous scientists and technologists who will be mobilized to attend to the consequent and inevitable problems that arise in a time of nuclear disaster.

Unfortunately there are no zero risks in the generation of nuclear energy. No matter how well-meaning re-assurances of safety are, the reality is that the risk is always there.

Furthermore, before establishing such a complex money guzzler as a nuclear power plant, Nigeria is obligated to consider seriously her past and present experiences in handling large sums of money that are meant for essential public services.

Finally, it is understandable that the citizens of a technologically impoverished country will have a hard time feeling safe with nuclear plants in their midst in view of the type of nuclear power accidents that occurred in Chernobyl and Fukushima.

Student: Thank you so much Prof. for your help.

Professor: You're welcome, and if you need additional help or a reading list for the subject we discussed, don't hesitate to come back.

Life expectancy in Nigeria

AS A RESULT of the brevity of human life here on earth and the desire of human beings to live healthy long lives, reflections on life expectancy should be of interest to all people.

Sometimes life expectancy is used as a synonym for **longevity**, but the latter really means "length of life or long life" as in, "your longevity in that job is outstanding", or "the pharmaceutical industry's dream is to discover and bottle genes for longevity".

Lifespan has also been unwittingly used as a synonym for life expectancy from which it differs significantly. Human lifespan is the maximum length of life that human beings can potentially attain. It is fixed through scientific research at about 100 years. This fixed maximum length of life is as true of Americans as it is of the British, French, Swedes, Japanese, Russians, and Nigerians or any other nationality.

Although it is inherent in all human beings to be able to live to about 100 years, only a very small number of people in any society reach this lifespan potential. This is mainly because human beings invariably die prematurely as a result of diseases, shootings, accidents or malnutrition. Consequently, if the prevalence of accidents and fatal diseases are high in a region then it is less likely that inhabitants of that region will live out their biologically endowed maximum lifespan potential.

Why is human lifespan fixed at about 100 years?

Although biologists do not yet know the answer to the above question, what appears certain is that lifespan is somehow genetically predetermined just as the length of human pregnancy is genetically fixed at about nine months, and the growth and maturation of a child are predetermined and always follow a predictable time scale.

The greatest support for genetic predetermination of lifespan comes from studying the lifespans of other species of animals. Each species is found to have a finite lifespan that is unique to that species.

For example, in a protected environment such as the zoo or laboratory, a rat lives for a maximum of about three years, a horse 46 years, an ape 50 years and, as I already mentioned human beings have a maximum lifespan potential of about 100 years. The species differences in lifespan are in line with the operation of genetic predetermination.

In contrast to lifespan, **life expectancy** is defined as the number of years that individuals born in the same year can hope to attain at birth if mortality at each age remains constant in the society in the future, or, it is the number of years of life remaining for any individual at a given age.

So if you are 50 years old and you were born and live in an area where the life expectancy is 70 years, your life expectancy is 20 years. Because life expectancy is a mathematical average, a person may die many years before or live many years after the life expectancy of his cohort. So unlike lifespan, life expectancy is not fixed.

Women have always outlived men since the prehistoric times to the present and as a result they have higher life expectancy than men. Consequently, life expectancy can be reported for males, for females and for the total population. For Nigerians in 2012 the corresponding values were: 49 years for men, 55 years for women and 52 years for the total population. Unless otherwise indicated I will henceforth use only life expectancy for the total population in this essay.

The reason(s) for the greater life expectancy for women when compared to men remains unclear in spite of the many studies that have been devoted to the subject. The theories that have been proposed include: First, men and women have differences in hormonal and immune status and response. Second, men are exposed more than women to specific risk factors such as alcohol consumption, cigarette smoking and war. Third, women are smaller in size than men and small body size has been found to increase lifespan in experimental animals like rodents. However, these suggestions have remained only as speculations.

Over the years life expectancy at birth has been increasing progressively in industrialized nations. In America it was 49.2 years in 1900 and 79 years in 2011. The corresponding data for Japan are 44 years in 1900 and 82 years in 2011.

The world Factbook reported in 2013 that the countries with the highest life expectancies in the world were Monaco 89.6 years, Japan 84.2 years, and Singapore 84.1 years. In the United States life expectancy was 79.5 years and she was ranked 43rd. In most advanced countries, however, life expectancy was about 81 years; in many developing countries it was about 51 years and in some it was below 50 years!

It is therefore not astonishing that life expectancy was reported to be 51.6 years in Nigeria in the year 2000 and 52.1 in 2012. In 2009, life expectancy in Nigeria was only 46.9 years making it one of the lowest among developing countries. This markedly low level was attributed mainly to the negative effect of HIV/AIDS disease which was rampant in Nigeria.

I believe that no one is particularly shocked that Nigeria has a low life expectancy because since political independence the country has remained an embodiment of the major factors that will shorten life expectancy. Unfortunately I do not see any reason to be optimistic that a substantial improvement of Nigeria's ranking in the world's life expectancy table is imminent.

So why is life expectancy so low in Nigeria?

Before I address this question let us first consider the fact that in the wild, animals die off long before they attain their maximum lifespan potential and only very few live until old age. However, improving and removing risk factors from their environment like housing them in the safety of the laboratory or the zoo increase the actual length of life that these animals attain.

The same is also true of humans. In advanced societies many people live long lives with little infirmity and they die mostly of old age because they have rid their societies of most elements deleterious to the survival of the individual.

However, in advanced societies there is still a small number of inevitable deaths from infant mortality due to the birth of defective and premature babies and a small number of deaths due to accidents and chronic diseases in adults. If they can successfully eliminate these deaths, advanced societies will improve the chances of their life expectancies approaching their biologically endowed maximum lifespan potential of 100 years.

Monaco appears to be already heading this way with her current life expectancy of almost 90 years.

It is now easy to see why life expectancy is so low in Nigeria. It is very low because Nigerians have failed to remove from their society those factors that are deleterious to human survival, and eventually those same factors cause Nigerians to die prematurely.

As was mentioned before the ability to live long may be built into the genetic code. In fact, in 2010 scientists at Boston University, USA even claimed to have found in humans unique genetic "signatures" that are strongly associated with long life.

So the genes they inherited from their parents may be the reason why the Nigerian twins, Mrs. Olukoya and Mrs. Ogunde were able to celebrate their 100th birthday recently in a country where the life expectancy is only about 52 years.

However, twin studies indicate that genes only contribute about 20 to 30 per cent of an individual's lifespan.

The consensus is that environmental factors and lifestyle have the biggest impact on life expectancy. This is good news because it means that it is within the power of individuals and any country to influence how long and how well their people live.

At the beginning of the 20th century infant mortality was very high around the world and life expectancy was correspondingly low. For example, as I mentioned earlier, in 1900 life expectancy at birth was only 49.2 years in the United States and 44 years in Japan, in keeping with worldwide high infant mortality.

The first enormous gains in life expectancy in industrialized nations occurred in the first half of the 20th century with the discovery of the "germ theory of disease".

This scientific breakthrough led to a new approach to preventive medicine as emphasis shifted to the importance of sanitation such as washing hands, protecting food from flies, improving water supply and proper sewage disposal.

These simple practices led to the control and eradication of many infectious and parasitic diseases.

Consequently mortality especially in infants and children markedly decreased resulting in a striking increase in national life expectancy in the industrialized world.

Similarly in Nigeria improved standards of hygiene, sanitary engineering, better nutrition, safe drinking water and the control and eradication of infectious diseases are bound to have positive effects on our national life expectancy. The level of scientific understanding that is required by these measures is no more arduous than that expected of beginning students in Nigerian medical and nursing schools. The technologies involved have been public knowledge for decades and are not beyond the capability of Nigeria.

Effective application of these preventive public health measures in Nigeria is obligatory for the elimination of needless premature deaths that will pave the way to the elongation of the life expectancy of Nigerians. The only barriers to implementing these simple measures are those erected by us.

Nonetheless, even when infant mortality is reduced or eliminated, there remains in every nation including Nigeria chronic diseases which if removed will have measurable positive effect on life expectancy. Therefore, another strategy that will increase the life expectancy of Nigerians is the prevention and control of these chronic diseases in adults.

To do this, good affordable healthcare is a necessity. In fact, the latter ranks very high if not the highest in many people's lists of the most desirable factors that have positive impact on life expectancy.

Good healthcare is expensive and among other things requires a good supply of doctors and other healthcare personnel. Unfortunately, at present medical service is poor and inefficient in Nigeria, and in rural areas it may be virtually non-existent due to a generalized shortage in the country of well-equipped hospitals and trained healthcare personnel including doctors.

This shortage is exacerbated by the migration of key healthcare personnel from Nigeria to greener pastures in other countries. For instance, many of the thousands of Nigerian medical doctors currently practicing in the United States and the United Kingdom received initial medical training in Nigeria.

In addition, most Nigerian healthcare personnel in the Western world refuse to return to Nigeria after completing their training overseas.

The shortage of key medical personal caused by these practices is tragic for Nigeria's healthcare system and the feasibility of using the country's medical services to increase life expectancy in Nigeria.

In line with the intuition that education plays a role in life expectancy, people with less than 12 years of education in the US were found to have lower life expectancy than their more educated cohorts.

This is probably because people with less education are likely to have: less income, poor housing, inadequate medical care, poor sanitation and poor nutrition, all of which have negative effects on life expectancy.

So Nigeria will also need to expand educational opportunities in the country in order to use education as a tool for increasing life expectancy.

From the above considerations alone it is clear that in any country there will be differences in life expectancy between regions and classes of people. Those that are in the upper socioeconomic class have a better chance of achieving higher life expectancy than those just struggling to meet their basic needs.

Since it is incumbent on any government to provide the foundation for a decent life for her citizens it seems trite that in 2015 I should be imploring Nigeria's governments to step up and position the country on the path of increasing the standard of living and life expectancy of ALL Nigerians by providing them: access to good healthcare, a culture of good sanitation, good roads, safe drinking water, affordable quality education, dependable electricity and other essential energy needs, adequate nutrition, environment with decreased crime, increased personal security, jobs for young people and an equitable society that abhors corruption by ALL people.

At the individual level Nigerians should aspire for a way of life that is likely to promote their life expectancy. For instance, they should eschew risky lifestyles such as smoking and excessive drinking of alcoholic beverages. Instead they should seek to: eat balanced diet, protect food from flies both in the market and at home, practice improved hygiene, engage in regular exercises, drink a lot of clean water throughout the day, wash their hands frequently, embark on continued personal educational enlightenment and do their bit to minimize pollution and protect the environment. Furthermore, they should avoid sexual practices that will likely increase the prevalence of HIV/AIDS disease, and they should support the government's efforts to prevent the spread of communicable diseases.

Finally, it is clear from these discussions that life expectancy is enhanced by many factors that are beneficial for the community, and that Nigeria is currently very far behind in the world life expectancy table.

I therefore propose that evaluation of life expectancy in various sectors of the society should be included as a measure of an administration's impact on the quality of life of Nigerians.

CPSIA information can be obtained at www.ICGtesting.com
Printed in the USA
BVOW05s2130081215

429789BV00001B/149/P